Chapter 1: Introduction to Acetylcholine

Overview of Acetylcholine and Its Role in the Nervous System

Acetylcholine (ACh) is a pivotal neurotransmitter that serves a multitude of functions in both the central and peripheral nervous systems. Discovered in the early 20th century, ACh was the first neurotransmitter to be identified, and its significance in synaptic transmission cannot be overstated. It plays an essential role in the transmission of signals across synapses, enabling communication between neurons, and facilitating numerous physiological processes.

ACh is synthesized from the precursors acetyl-CoA and choline through the action of the enzyme choline acetyltransferase. Once synthesized, acetylcholine is stored in vesicles at the presynaptic terminals of neurons. Upon stimulation, these vesicles fuse with the presynaptic membrane, releasing ACh into the synaptic cleft. This process is crucial for neuronal signaling, affecting everything from muscle contraction to cognitive function.

Importance in Neurotransmission and Muscle Activation

Acetylcholine's role in neurotransmission is particularly evident at the neuromuscular junction, where it mediates the transmission of impulses from motor neurons to skeletal muscles. When ACh is released from motor neurons, it binds to nicotinic acetylcholine receptors on muscle fibers, leading to depolarization of the muscle membrane and subsequent muscle contraction. This process is fundamental for voluntary movements, as well as reflex actions that enable us to respond quickly to stimuli.

In the central nervous system, ACh is involved in several critical functions, including attention, learning, and memory. It modulates synaptic plasticity, which is vital for memory formation and cognitive processes. Dysregulation of acetylcholine signaling has been implicated in various neurological disorders, including Alzheimer's disease, where decreased levels of ACh are associated with cognitive decline.

Moreover, ACh plays a significant role in the autonomic nervous system, influencing both the sympathetic and parasympathetic branches. It is the primary neurotransmitter of the parasympathetic nervous system, promoting rest and digest responses, such as lowering heart rate and enhancing digestion. Conversely, its role in the sympathetic nervous system, while less direct, is also important for the overall balance of autonomic functions.

Summary

Understanding acetylcholine's multifaceted roles in neurotransmission and muscle activation provides a foundation for exploring more complex interactions and implications in health and disease. The regulation of acetylcholine levels, particularly through the inhibition of acetylcholinesterase (AChE), presents significant therapeutic opportunities for enhancing cognitive function and addressing various neurological conditions. As we delve deeper into the mechanisms of acetylcholine action and regulation, the subsequent chapters will elucidate the intricate systems and therapeutic potentials that revolve around this essential neurotransmitter.

Chapter 2: The Acetylcholine Receptor System

Types of Acetylcholine Receptors: Nicotinic and Muscarinic

Acetylcholine exerts its effects through two main types of receptors: nicotinic and muscarinic receptors. Each type of receptor has distinct structures, functions, and physiological roles, reflecting the diverse actions of acetylcholine in the body.

Nicotinic Acetylcholine Receptors (nAChRs)

Nicotinic receptors are ligand-gated ion channels found at the neuromuscular junctions, autonomic ganglia, and in the central nervous system. These receptors are named after their activation by nicotine, a potent alkaloid found in tobacco. When acetylcholine binds to nicotinic receptors, it induces a conformational change that opens the channel, allowing sodium (Na^+) ions to flow into the cell and potassium (K^+) ions to exit. This ion flux leads to depolarization of the cell membrane, generating an excitatory postsynaptic potential.

Nicotinic receptors play a crucial role in muscle contraction, as they mediate the transmission of signals from motor neurons to skeletal muscle fibers. In the autonomic nervous system, nAChRs are involved in transmitting signals from preganglionic to postganglionic neurons, affecting both the sympathetic and parasympathetic pathways.

Nicotinic receptors can be further classified based on their subunit composition. The most studied are the muscle-type receptors, which consist of two α1, one β1, one δ, and one ε subunit, and the neuronal-type receptors, which can have various combinations of α and β subunits. This diversity allows for different functional properties and pharmacological profiles.

Muscarinic Acetylcholine Receptors (mAChRs)

Muscarinic receptors, named after their activation by muscarine, a compound derived from the mushroom Amanita muscaria, are G protein-coupled receptors found throughout the body. These receptors are primarily located in the heart, smooth muscles, and glandular tissues. There are five subtypes of muscarinic receptors, designated M1 through M5, each with distinct functions and tissue distributions.

Upon binding of acetylcholine, muscarinic receptors activate intracellular signaling pathways through G proteins, leading to various cellular responses. For example, the M2 subtype is predominantly found in the heart, where its activation results in decreased heart rate and reduced force of contraction. Conversely, the M3 subtype, found in smooth muscle and glands, mediates contraction of smooth muscle and secretion of glandular fluids.

The diversity of muscarinic receptor subtypes allows acetylcholine to elicit a range of physiological responses, from regulating heart rate to modulating glandular secretions and controlling smooth muscle contractions.

Mechanisms of Action and Physiological Effects

The mechanisms of action for nicotinic and muscarinic receptors illustrate the complexity of acetylcholine's role in physiological processes.

Mechanism of Action

1. **Nicotinic Receptors**: As mentioned earlier, nicotinic receptors are ionotropic receptors that respond to acetylcholine by opening ion channels. The rapid influx of Na^+ ions depolarizes the postsynaptic membrane, leading to excitatory effects in muscle cells and neurons. This mechanism is critical for quick responses, such as muscle contraction and neuronal signaling.

2. **Muscarinic Receptors**: Muscarinic receptors, being metabotropic, initiate slower and more prolonged responses. Upon activation by acetylcholine, these receptors interact with G proteins to modulate intracellular signaling pathways. For instance, M2 receptors inhibit adenylate cyclase activity, reducing cyclic AMP (cAMP) levels, while M3 receptors stimulate phospholipase C, leading to the production of inositol trisphosphate (IP3) and diacylglycerol (DAG), which promote intracellular calcium release and muscle contraction.

Physiological Effects

The physiological effects of acetylcholine, mediated by its receptors, encompass a wide range of functions:

- **Muscle Contraction**: Activation of nicotinic receptors at the neuromuscular junction is essential for voluntary muscle movement, as it initiates the excitation-contraction coupling process.
- **Cognitive Functions**: In the brain, acetylcholine is crucial for attention, learning, and memory. Its action on nicotinic receptors enhances neurotransmission and synaptic plasticity, which are vital for cognitive processes.
- **Autonomic Functions**: Through muscarinic receptors, acetylcholine regulates autonomic functions such as heart rate, digestion, and glandular secretion. The parasympathetic activation of mAChRs leads to a "rest and digest" state, promoting relaxation and energy conservation.
- **Pain Modulation**: Acetylcholine also plays a role in modulating pain pathways, where its action on both nicotinic and muscarinic receptors can influence pain perception and response.

Summary

Understanding the acetylcholine receptor system is crucial for mastering the complexities of acetylcholine's actions in the body. The distinct roles of nicotinic and muscarinic receptors, along with their mechanisms of action, highlight the importance of acetylcholine in neurotransmission, muscle activation, and a wide array of physiological functions. As we move forward in this book, we will delve deeper into the synthesis and release of acetylcholine, exploring how its regulation can be manipulated through acetylcholinesterase inhibition to harness therapeutic benefits.

Chapter 3: Synthesis and Release of Acetylcholine

Biosynthesis Pathways

The synthesis of acetylcholine (ACh) is a finely tuned biochemical process that occurs in cholinergic neurons. This process primarily involves two key precursors: acetyl-CoA and choline.

1. Acetyl-CoA Production

Acetyl-CoA, the acetyl group donor for ACh synthesis, is derived from several metabolic pathways. It can be produced through the breakdown of carbohydrates (via glycolysis), fats (through β-oxidation), and proteins (from amino acids). Once synthesized, acetyl-CoA is transported to the cytoplasm of the neuron.

2. Choline Uptake

Choline is an essential nutrient that can be obtained from dietary sources such as meat, fish, eggs, and certain plant foods. Once choline is available, it enters the neuron through high-affinity choline transporters located in the presynaptic membrane. This transporter, known as the sodium-dependent choline transporter (CHT), plays a critical role in ensuring adequate choline levels for ACh synthesis.

3. ACh Synthesis

Once inside the neuron, acetyl-CoA and choline undergo a condensation reaction catalyzed by the enzyme choline acetyltransferase (ChAT). This enzyme facilitates the transfer of the acetyl group from acetyl-CoA to choline, resulting in the formation of acetylcholine. This synthesis occurs primarily in the cytoplasm of cholinergic neurons.

4. Storage in Vesicles

After synthesis, ACh is packaged into synaptic vesicles by the vesicular acetylcholine transporter (VAChT). This packaging is crucial for the storage of ACh until it is needed for neurotransmission. The vesicles protect ACh from degradation and help maintain a reserve for rapid release during synaptic transmission.

Release Mechanisms in Synaptic Transmission

The release of acetylcholine from presynaptic neurons is a highly regulated process that occurs at the neuromuscular junction and within the central nervous system. This process involves several key steps:

1. Action Potential Arrival

The release of ACh is triggered by the arrival of an action potential at the presynaptic terminal. The depolarization caused by the action potential opens voltage-gated calcium (Ca^{2+}) channels in the presynaptic membrane.

2. Calcium Influx

As a result of the action potential, calcium ions rush into the neuron through these opened channels. The increase in intracellular calcium concentration is a critical signal that initiates the release of ACh.

3. Vesicle Fusion

The influx of calcium ions activates proteins known as synaptotagmins, which facilitate the fusion of synaptic vesicles containing ACh with the presynaptic membrane. This process is mediated by SNARE proteins, which are essential for the docking and fusion of vesicles.

4. Exocytosis of Acetylcholine

Once the vesicles fuse with the membrane, acetylcholine is released into the synaptic cleft through a process known as exocytosis. This release is a rapid process, allowing for quick signaling between neurons or between neurons and muscle cells.

5. Binding to Receptors

After its release, ACh diffuses across the synaptic cleft and binds to nicotinic or muscarinic receptors on the postsynaptic membrane, depending on the type of synapse. This binding initiates a cascade of events, leading to either muscle contraction or modulation of neurotransmission in neurons.

Summary

The synthesis and release of acetylcholine are critical processes that underpin its role as a neurotransmitter in the nervous system. Through the intricate pathways of acetyl-CoA and choline metabolism, along with the precise mechanisms of vesicle release and receptor activation, ACh plays a fundamental role in muscle activation, cognitive function, and numerous autonomic processes. Understanding these pathways is essential for exploring how modulation of acetylcholine levels—particularly through the inhibition of acetylcholinesterase—can have therapeutic benefits. The subsequent chapters will delve into the degradation of acetylcholine and the implications of its regulation in health and disease.

Chapter 4: Acetylcholinesterase: The Enzyme of Degradation

Function and Significance of Acetylcholinesterase

Acetylcholinesterase (AChE) is a vital enzyme in the nervous system responsible for the hydrolysis of acetylcholine, a neurotransmitter essential for neurotransmission and muscle activation. Found predominantly at synapses and neuromuscular junctions, AChE serves to terminate the action of acetylcholine after its release, ensuring precise control over synaptic transmission and muscle contraction.

Mechanism of Action

The enzymatic activity of AChE involves the breakdown of acetylcholine into acetate and choline, a process that occurs within milliseconds after acetylcholine binds to its receptors. AChE catalyzes this reaction through a two-step mechanism:

1. **Acetylation**: Acetylcholine binds to the active site of AChE, leading to the formation of an acetyl-enzyme complex. This intermediate state temporarily holds the acetyl group of acetylcholine, facilitating its cleavage.
2. **Hydrolysis**: Water molecules are then introduced into the reaction, causing the release of acetate and regenerating the enzyme. This hydrolysis not only terminates the action of acetylcholine but also recycles choline back into the neuron for re-synthesis of acetylcholine.

AChE is crucial for maintaining the dynamic equilibrium of neurotransmission. By swiftly degrading acetylcholine, it prevents prolonged stimulation of receptors, which could lead to continuous muscle contraction or excessive neuronal firing, both of which can be detrimental to physiological function.

Role in Regulating Acetylcholine Levels

The regulation of acetylcholine levels in the synaptic cleft is critical for proper nervous system function. AChE plays a central role in this regulation, ensuring that acetylcholine is available for neurotransmission when needed, but also swiftly removed when its action is no longer required.

1. Maintaining Homeostasis

By controlling acetylcholine concentration, AChE contributes to synaptic homeostasis. This balance is essential for effective communication between neurons and between motor neurons and muscle fibers. Dysregulation of AChE activity can lead to pathological conditions; for example, reduced AChE activity can result in excessive acetylcholine accumulation, leading to symptoms such as muscle spasms, paralysis, or even respiratory failure.

2. Implications for Disease

In various neurological disorders, the activity of AChE is altered, leading to imbalances in acetylcholine levels. For instance, in Alzheimer's disease, the degeneration of cholinergic neurons results in decreased acetylcholine availability, contributing to cognitive decline. In contrast, conditions like myasthenia gravis involve the production of antibodies against nicotinic receptors, leading to compensatory increases in acetylcholine levels. In both scenarios, understanding AChE's role becomes crucial for developing therapeutic strategies aimed at modulating acetylcholine levels.

3. Pharmacological Target

The critical role of AChE in regulating acetylcholine levels has made it a prime target for pharmacological interventions. Acetylcholinesterase inhibitors (AChEIs) are used in clinical practice to increase acetylcholine availability in synapses, enhancing neurotransmission. This therapeutic approach has shown significant benefits in conditions such as Alzheimer's disease and myasthenia gravis, highlighting the importance of AChE in both health and disease.

Summary

Acetylcholinesterase is an essential enzyme that regulates the action of acetylcholine, playing a pivotal role in neurotransmission and muscle activation. Through its rapid hydrolysis of acetylcholine, AChE maintains synaptic homeostasis, prevents overstimulation, and contributes to the overall balance of neurotransmitter levels in the nervous system. Understanding the function and significance of AChE not only deepens our knowledge of cholinergic signaling but also underscores the potential for therapeutic interventions that target this enzyme. As we progress in this book, we will explore the pharmacology of AChE inhibitors and their clinical applications in various neurological disorders.

Chapter 5: Pharmacology of Acetylcholinesterase Inhibitors

Overview of Acetylcholinesterase Inhibitors (AChEIs)

Acetylcholinesterase inhibitors (AChEIs) are a class of compounds that inhibit the activity of the enzyme acetylcholinesterase, thereby prolonging the action of acetylcholine in the synaptic cleft. This pharmacological approach enhances cholinergic transmission and has been extensively studied for its therapeutic applications, particularly in neurodegenerative diseases and other conditions characterized by impaired cholinergic function.

The inhibition of AChE leads to increased levels of acetylcholine, allowing for enhanced stimulation of both nicotinic and muscarinic receptors. AChEIs can be classified into two main categories based on their mechanism of action: reversible and irreversible inhibitors.

Therapeutic Uses and Mechanisms of Action

1. Reversible AChE Inhibitors

Reversible AChE inhibitors bind to the active site of acetylcholinesterase and form a temporary complex, allowing for the restoration of enzyme activity over time. These inhibitors are commonly used in clinical settings and include:

- **Donepezil**: Used primarily for the treatment of Alzheimer's disease, donepezil selectively inhibits AChE in the brain, resulting in increased levels of acetylcholine. It has been shown to improve cognitive function and slow disease progression in some patients.
- **Rivastigmine**: Another AChEI used for Alzheimer's and Parkinson's disease-related dementia, rivastigmine is unique as it inhibits both AChE and butyrylcholinesterase (BuChE), an enzyme that also breaks down acetylcholine. This dual inhibition enhances its therapeutic effects.
- **Galantamine**: In addition to its AChE inhibition, galantamine enhances the release of acetylcholine by acting on nicotinic receptors. This makes it particularly effective for improving memory and cognition in patients with Alzheimer's disease.

The mechanisms of these reversible AChEIs primarily focus on enhancing synaptic transmission and improving communication between neurons, thereby alleviating some symptoms of cognitive decline.

2. Irreversible AChE Inhibitors

Irreversible AChE inhibitors bind covalently to the active site of acetylcholinesterase, permanently inactivating the enzyme. This class of inhibitors has limited therapeutic application but is significant in toxicology and pharmacology. Examples include:

- **Organophosphates**: Commonly used in pesticides and as nerve agents, these compounds inhibit AChE irreversibly, leading to excessive accumulation of acetylcholine. The resultant cholinergic crisis can cause severe toxicity and requires immediate medical intervention.
- **Nerve Agents**: Compounds such as sarin and VX function as potent irreversible AChE inhibitors, leading to overstimulation of cholinergic receptors and potentially fatal outcomes if not treated promptly with antidotes like atropine or pralidoxime.

While irreversible AChE inhibitors are not typically used therapeutically, their study provides valuable insights into the mechanisms of cholinergic signaling and the potential consequences of AChE inhibition.

Clinical Implications of AChEIs

The use of AChE inhibitors in clinical practice highlights their importance in treating conditions associated with cholinergic deficits. The enhancement of acetylcholine levels has demonstrated efficacy in improving cognitive function, memory, and overall quality of life for patients with Alzheimer's disease and other dementias. However, the clinical use of AChEIs is not without challenges.

1. Efficacy and Limitations

While AChEIs can provide symptomatic relief, they do not stop the underlying neurodegenerative processes associated with diseases like Alzheimer's. Therefore, their effects are often modest and may vary among individuals. Additionally, the long-term effectiveness of these treatments can diminish as the disease progresses.

2. Side Effects and Risks

The increased levels of acetylcholine resulting from AChEI therapy can lead to side effects, including nausea, diarrhea, muscle cramps, and increased salivation. In some cases, patients may experience bradycardia (slow heart rate) or fainting due to enhanced parasympathetic activity. Monitoring and dose adjustments are often necessary to mitigate these risks.

3. Emerging Research

Current research is exploring new AChEI compounds and alternative strategies for cholinergic enhancement. Some studies are investigating the combination of AChEIs with other therapeutic agents, such as NMDA receptor antagonists, to address the multifactorial nature of neurodegenerative diseases.

Summary

Acetylcholinesterase inhibitors represent a critical pharmacological strategy for enhancing cholinergic transmission and treating conditions characterized by acetylcholine deficits. Through reversible and irreversible inhibition of AChE, these compounds can significantly impact cognitive function and overall neurological health. However, their clinical application necessitates careful consideration of efficacy, side effects, and long-term implications. As research continues to evolve, AChEIs may play an even more prominent role in managing neurodegenerative diseases and other related disorders, paving the way for innovative therapeutic approaches.

Chapter 6: Clinical Applications of AChEIs

Treatment of Alzheimer's Disease

Alzheimer's disease (AD) is a progressive neurodegenerative disorder characterized by cognitive decline, memory loss, and changes in behavior. It is the most common cause of dementia, affecting millions of people worldwide. The pathophysiology of Alzheimer's disease involves a decrease in acetylcholine levels due to the degeneration of cholinergic neurons in the brain, particularly in regions crucial for learning and memory, such as the hippocampus and cortex.

1. Role of AChE Inhibitors in AD

Acetylcholinesterase inhibitors (AChEIs) have emerged as a cornerstone of pharmacological treatment for Alzheimer's disease. By inhibiting the enzyme that breaks down acetylcholine, these drugs aim to increase the availability of acetylcholine in the synaptic cleft, thereby enhancing cholinergic transmission and potentially improving cognitive function.

Key AChEIs Used in Alzheimer's Disease:

- **Donepezil**: Approved for mild to moderate Alzheimer's, donepezil selectively inhibits AChE in the brain, leading to improved cognitive function and activities of daily living in some patients.
- **Rivastigmine**: Effective for both Alzheimer's disease and Parkinson's disease dementia, rivastigmine's dual inhibition of AChE and butyrylcholinesterase provides broader cholinergic enhancement.
- **Galantamine**: This AChEI not only inhibits AChE but also modulates nicotinic receptors, further enhancing its cognitive benefits.

Clinical studies have shown that these AChEIs can produce modest improvements in cognition and function, particularly in the early to moderate stages of the disease. However, the benefits are generally symptomatic rather than curative, emphasizing the need for ongoing research into more effective treatments.

Other Neurological Disorders and Their Management

AChEIs are not only beneficial in Alzheimer's disease but also play a significant role in the management of other neurological disorders characterized by cholinergic dysfunction.

1. Myasthenia Gravis

Myasthenia gravis (MG) is an autoimmune disorder that affects the neuromuscular junction, leading to weakness and fatigue of voluntary muscles. The condition is caused by the production of antibodies that block or destroy nicotinic receptors at the neuromuscular junction.

Treatment with AChEIs:

Pyridostigmine

2. Dementia with Lewy Bodies

Dementia with Lewy bodies (DLB) shares clinical features with both Alzheimer's disease and Parkinson's disease. Patients often experience cognitive fluctuations, visual hallucinations, and parkinsonism. The cholinergic system is also implicated in DLB.

AChEI Use in DLB:

Rivastigmine

3. Postoperative Cognitive Dysfunction

Postoperative cognitive dysfunction (POCD) is a decline in cognitive function that can occur after surgery, particularly in elderly patients. The underlying mechanisms are not fully understood, but there is evidence to suggest that cholinergic deficits may contribute to this condition.

Potential AChEI Role: Research is ongoing to evaluate the efficacy of AChEIs in preventing or treating POCD. The goal is to enhance cholinergic function and potentially mitigate cognitive decline following surgical procedures.

Summary

Acetylcholinesterase inhibitors play a crucial role in the clinical management of Alzheimer's disease and other neurological disorders characterized by cholinergic deficits. By enhancing acetylcholine levels, AChEIs can improve cognitive function, muscle strength, and overall quality of life for patients. Despite their benefits, it is essential to recognize that AChEIs primarily provide symptomatic relief and do not halt the progression of neurodegenerative diseases. Ongoing research is necessary to explore new AChEIs, combination therapies, and alternative approaches to optimize patient outcomes in cholinergic dysfunctions.

Chapter 7: Mechanisms of AChE Inhibition
How AChE Inhibitors Block Acetylcholinesterase

Acetylcholinesterase inhibitors (AChEIs) play a crucial role in enhancing cholinergic transmission by blocking the activity of the enzyme acetylcholinesterase (AChE). Understanding the mechanisms by which these inhibitors function is essential for both therapeutic applications and the development of new drugs targeting cholinergic systems.

1. Mechanisms of Inhibition

AChE inhibitors can be classified into two broad categories based on their interaction with the AChE enzyme: reversible and irreversible inhibitors. Each type has distinct mechanisms of action that dictate their pharmacological effects and therapeutic uses.

Reversible Inhibitors

- **Competitive Inhibition**: The inhibitor binds to the active site of AChE, preventing acetylcholine from accessing the enzyme. By inhibiting AChE activity, these drugs prolong the presence of acetylcholine in the synaptic cleft, enhancing cholinergic transmission. This is the mechanism utilized by drugs such as donepezil and rivastigmine.
- **Non-competitive Inhibition**: Some reversible inhibitors can bind to sites other than the active site (allosteric sites) on the enzyme. This binding alters the enzyme's conformation, reducing its activity regardless of whether acetylcholine is present. This mechanism allows for an increase in acetylcholine levels without direct competition.

Irreversible Inhibitors

Phosphorylation

Differences Between Reversible and Irreversible Inhibitors

The differences between reversible and irreversible AChE inhibitors have important implications for their therapeutic uses, potential side effects, and overall safety profiles.

1. Duration of Action

- **Reversible Inhibitors**: These compounds have a shorter duration of action, allowing for more controlled and flexible dosing. Their effects can be quickly reversed if necessary, making them suitable for chronic conditions like Alzheimer's disease, where modulation of cholinergic signaling is needed without permanent changes.
- **Irreversible Inhibitors**: These have a prolonged duration of action, as they permanently inactivate AChE. This can lead to toxic effects, particularly in the case of accidental exposure to organophosphates or nerve agents. The long-lasting inhibition necessitates immediate medical intervention to restore AChE activity.

2. Clinical Applications

- **Reversible Inhibitors**: Primarily used in the treatment of cognitive disorders, such as Alzheimer's disease and myasthenia gravis. Their ability to enhance synaptic transmission in a controlled manner makes them suitable for managing symptoms without overwhelming cholinergic systems.
- **Irreversible Inhibitors**: While not used therapeutically, they serve critical roles in toxicology and chemical warfare. Understanding their mechanisms helps in developing antidotes and treatment strategies for poisoning, such as the use of atropine to counteract the effects of nerve agents.

3. Side Effects and Toxicity

- **Reversible Inhibitors**: Generally, these have a more favorable safety profile, but they can still cause side effects related to cholinergic overstimulation, including gastrointestinal symptoms, muscle cramps, and increased salivation.
- **Irreversible Inhibitors**: These pose significant risks of toxicity due to prolonged acetylcholine accumulation. Symptoms of poisoning can include muscle paralysis, respiratory failure, and potentially fatal outcomes without prompt treatment.

Summary

Understanding the mechanisms by which acetylcholinesterase inhibitors operate—both reversible and irreversible—is essential for their effective use in clinical practice. Reversible inhibitors provide valuable therapeutic options for conditions characterized by cholinergic deficits, while irreversible inhibitors underscore the importance of careful handling and understanding in the context of toxicology. As research progresses, the development of new AChEIs that maximize therapeutic benefits while minimizing side effects will continue to be a focus in pharmacology and neurobiology.

Chapter 8: Side Effects and Toxicity of AChE Inhibitors

Common Side Effects and Risk Factors

Acetylcholinesterase inhibitors (AChEIs), while beneficial for enhancing cholinergic transmission, are not without their risks and side effects. The increase in acetylcholine levels in the synaptic cleft, although therapeutic in some contexts, can lead to overstimulation of cholinergic receptors and result in various adverse effects. Understanding these side effects is essential for clinicians and patients alike.

1. Common Side Effects

The side effects of AChE inhibitors can vary based on the specific drug used, the dosage, and individual patient factors. However, some common side effects include:

- **Gastrointestinal Symptoms**: Many patients experience nausea, vomiting, diarrhea, and abdominal cramps. These symptoms arise from increased cholinergic activity in the gastrointestinal tract, which can lead to heightened secretory activity and increased motility.
- **Muscle Cramps and Twitching**: Due to enhanced neuromuscular transmission, patients may experience muscle cramps, fasciculations (muscle twitching), and general muscle stiffness.
- **Bradycardia**: AChE inhibitors can cause a decrease in heart rate, leading to bradycardia. This occurs due to increased activation of muscarinic receptors in the heart, which can potentially lead to dizziness or fainting in some patients.
- **Increased Salivation and Sweating**: Cholinergic overstimulation can lead to increased salivary secretion and perspiration, which can be uncomfortable for patients.
- **Visual Disturbances**: Some patients may experience blurred vision or difficulty focusing, primarily due to increased activity at the muscarinic receptors in the eye.

2. Risk Factors for Adverse Effects

Certain factors can increase the likelihood of experiencing side effects from AChEIs:

- **Age**: Older patients may be more susceptible to side effects due to decreased metabolic function and the presence of comorbidities.
- **Co-existing Medical Conditions**: Patients with cardiac issues, gastrointestinal disorders, or respiratory conditions may face higher risks for adverse effects when treated with AChEIs.
- **Drug Interactions**: AChEIs can interact with other medications, especially those that also influence cholinergic activity, leading to increased side effects. For example, the combination of AChEIs with anticholinergic drugs can reduce the efficacy of treatment and may cause additional side effects.

Long-term Implications of AChE Inhibition

The long-term use of AChE inhibitors, while effective in managing symptoms of diseases like Alzheimer's and myasthenia gravis, raises concerns about potential cumulative effects and long-term safety.

1. Tolerance Development

Some patients may develop tolerance to the effects of AChE inhibitors over time, leading to diminished therapeutic benefits. This phenomenon can necessitate dose adjustments or switching to alternative treatments.

2. Potential for Toxicity

- **Cholinergic Crisis**: Prolonged use or overdose of AChE inhibitors can lead to a cholinergic crisis, characterized by excessive stimulation of cholinergic receptors. Symptoms may include severe muscle weakness, respiratory distress, convulsions, and even coma. This condition is a medical emergency that requires immediate intervention.
- **Neurological Effects**: Chronic overstimulation of cholinergic pathways may lead to long-term neurological effects. While research is still ongoing, concerns exist regarding potential impacts on cognitive function or exacerbation of neurodegenerative processes in vulnerable populations.
- **Withdrawal Effects**: Discontinuation of AChEIs after prolonged use may lead to withdrawal symptoms, including a sudden decrease in cholinergic activity, which can exacerbate cognitive decline or muscular weakness in conditions like myasthenia gravis.

Summary

While acetylcholinesterase inhibitors are valuable tools in managing cholinergic dysfunction, they carry the potential for side effects and toxicity. Common adverse effects include gastrointestinal disturbances, muscle cramps, bradycardia, and increased secretions. Understanding the risk factors that contribute to these side effects is crucial for optimizing treatment and ensuring patient safety. Long-term use of AChEIs requires careful monitoring for potential tolerance, toxicity, and withdrawal effects. As research continues, better strategies may emerge to mitigate these risks while maximizing the therapeutic benefits of AChE inhibitors.

Chapter 9: Natural AChE Inhibitors: A Herbal Perspective

Overview of Natural Compounds and Their Effects

In addition to synthetic acetylcholinesterase inhibitors (AChEIs) used in clinical settings, various natural compounds derived from plants have shown potential as AChE inhibitors. These natural AChEIs can offer therapeutic benefits with potentially fewer side effects than their synthetic counterparts. This chapter will explore the mechanisms of action of these compounds, their sources, and their implications for health and wellness.

Mechanisms of Action

Natural AChE inhibitors typically exert their effects through mechanisms similar to those of synthetic AChEIs, primarily by binding to the active site of acetylcholinesterase and preventing it from hydrolyzing acetylcholine. The effectiveness of these natural compounds often relies on their chemical structure, which allows them to compete with acetylcholine or to bind covalently to the enzyme.

1. **Competitive Inhibition**: Many plant-derived AChEIs act by competing with acetylcholine for the active site of the enzyme. This action increases the concentration of acetylcholine available in the synaptic cleft, thus enhancing cholinergic transmission.
2. **Allosteric Modulation**: Some natural AChEIs may bind to sites other than the active site (allosteric sites) on the enzyme, leading to conformational changes that reduce enzymatic activity without directly competing with acetylcholine.

Examples of Plants Used as AChE Inhibitors

Numerous plants have been studied for their potential AChE inhibitory properties. Below are some notable examples:

1. Ginkgo Biloba

Ginkgo biloba is one of the most researched herbal supplements. Its extract contains flavonoids and terpenoids, which have been shown to possess AChE inhibitory activity. Research indicates that ginkgo may improve cognitive function in patients with dementia, potentially through its effects on cholinergic signaling.

2. Huperzia serrata

Huperzia serrata, a species of clubmoss, contains huperzine A, a potent reversible AChEI. This compound has gained attention for its ability to enhance memory and cognitive function. Clinical studies suggest that huperzine A may be effective in treating Alzheimer's disease and improving memory performance in healthy individuals.

3. Rosemary (Rosmarinus officinalis)

Rosemary is not only a culinary herb but also has been shown to have medicinal properties. Compounds such as rosmarinic acid and other phenolic compounds in rosemary exhibit AChE inhibitory effects. The herb is believed to enhance cognitive function and memory, making it a popular choice in traditional medicine.

4. Bacopa monnieri

Bacopa monnieri, also known as Brahmi, is an Ayurvedic herb that has been traditionally used to enhance memory and cognitive performance. Studies have demonstrated that bacopa extracts can inhibit AChE, leading to increased acetylcholine levels and improved cognitive function.

5. Sage (Salvia officinalis)

Sage is another herb with historical use in enhancing memory and cognitive function. Research has identified several compounds in sage, such as carnosic acid, that possess AChE inhibitory activity, potentially contributing to its neuroprotective effects.

6. Ashwagandha (Withania somnifera)

Ashwagandha is an adaptogenic herb known for its stress-relieving properties. Some studies suggest that it may also exhibit AChE inhibitory activity, contributing to improved cognitive function and memory.

Implications for Health and Wellness

The potential of natural AChE inhibitors presents exciting opportunities for health and wellness. These compounds may offer alternative or adjunct therapies for cognitive decline and neurodegenerative diseases. However, it is essential to consider the following:

- **Quality and Standardization**: The efficacy of herbal supplements can vary widely based on factors such as plant species, preparation methods, and dosages. Standardization of extracts is crucial for ensuring consistent therapeutic effects.
- **Interactions with Conventional Treatments**: Patients considering natural AChEIs should be aware of possible interactions with prescribed medications. Consulting with healthcare providers before starting any new herbal regimen is essential.
- **Further Research**: While preliminary studies show promise, more extensive clinical trials are necessary to establish the safety, efficacy, and optimal dosages of these natural AChEIs in treating conditions related to cholinergic dysfunction.

Summary

Natural acetylcholinesterase inhibitors derived from plants represent a promising area of research in enhancing cognitive function and managing neurodegenerative diseases. With their potential for fewer side effects compared to synthetic AChEIs, these herbal compounds could play a valuable role in holistic approaches to cognitive health. As interest in natural remedies continues to grow, understanding their mechanisms and implications will be critical for integrating these therapies into modern healthcare practices.

Chapter 10: Acetylcholine and Learning & Memory

The Role of Acetylcholine in Cognitive Functions

Acetylcholine (ACh) is a critical neurotransmitter involved in various cognitive processes, particularly learning and memory. Its influence on these functions is well-documented, and understanding the role of ACh can provide insights into how cognitive impairments occur and how they can be mitigated.

1. Mechanisms of Action in the Brain

Acetylcholine acts primarily within the central nervous system (CNS), where it is involved in modulating attention, arousal, and the encoding and retrieval of memories. Key areas in the brain where ACh exerts its effects include:

- **Hippocampus**: This region is essential for forming new memories. Cholinergic neurons originating from the basal forebrain project to the hippocampus, where they facilitate synaptic plasticity, a critical mechanism for learning and memory. Increased ACh levels enhance long-term potentiation (LTP), which strengthens synapses and improves memory formation.
- **Cortex**: Acetylcholine plays a role in attention and sensory processing. It enhances signal-to-noise ratios in neural circuits, allowing for better discrimination of relevant stimuli from background noise. This function is vital for effective learning as it helps prioritize important information.
- **Amygdala**: Involved in emotional learning and memory, the amygdala's cholinergic modulation influences how emotional contexts affect memory encoding and retrieval. ACh can enhance the memory of emotionally charged events, impacting how individuals learn from experiences.

Implications of AChE Inhibition on Memory Enhancement

Given the critical role of acetylcholine in cognitive processes, acetylcholinesterase inhibitors (AChEIs) are of particular interest for their potential to enhance memory and learning. By preventing the breakdown of acetylcholine, these inhibitors increase its availability in the synaptic cleft, leading to improved cholinergic signaling.

1. Enhancing Cognitive Function in Alzheimer's Disease

AChEIs are commonly prescribed to patients with Alzheimer's disease, a condition characterized by significant cholinergic deficits. Studies have demonstrated that these medications can lead to modest improvements in cognitive function, particularly in memory and daily living activities. The improvement is often measured using standardized cognitive assessments, showing that AChEIs can help slow the cognitive decline associated with the disease.

2. Potential Benefits in Healthy Individuals

Research has also explored the effects of AChEIs in healthy individuals, particularly concerning memory enhancement. Some studies suggest that AChEIs can improve memory performance and cognitive flexibility, especially in tasks that require attention and working memory. While the effects are generally modest, they indicate that enhancing cholinergic signaling could be a viable strategy for boosting cognitive performance in certain contexts.

Research and Future Directions

Ongoing research is focusing on better understanding the relationship between acetylcholine, learning, and memory. Areas of exploration include:

- **Neurogenesis**: There is emerging evidence that ACh may play a role in promoting neurogenesis (the formation of new neurons) in the hippocampus, suggesting that enhancing cholinergic activity could contribute to cognitive resilience and recovery.
- **Synaptic Plasticity**: Further studies are investigating how ACh influences synaptic plasticity at different stages of learning and memory consolidation. Understanding these mechanisms could lead to targeted therapies that enhance memory formation in both healthy and cognitively impaired populations.
- **Combination Therapies**: Research is also exploring the efficacy of combining AChEIs with other cognitive enhancers, such as NMDA receptor antagonists, to maximize their therapeutic benefits and address the multifactorial nature of cognitive decline.

Summary

Acetylcholine plays a vital role in learning and memory, affecting various cognitive processes through its action in critical brain regions. The use of acetylcholinesterase inhibitors to enhance cholinergic signaling has demonstrated potential in improving cognitive function, particularly in individuals with Alzheimer's disease. Ongoing research continues to explore the mechanisms of ACh in learning and memory, paving the way for innovative therapeutic approaches to support cognitive health. As our understanding of acetylcholine's role expands, so too does the potential for harnessing its effects to enhance memory and learning across diverse populations.

Chapter 11: Acetylcholine in Muscle Function

Neuromuscular Junction and Muscle Contraction

Acetylcholine (ACh) plays a pivotal role in muscle function, particularly at the neuromuscular junction (NMJ), where motor neurons communicate with skeletal muscle fibers. The NMJ is a specialized synapse that facilitates the transmission of electrical signals from the nerve to the muscle, enabling voluntary movements.

1. Structure of the Neuromuscular Junction

The NMJ consists of the presynaptic terminal of a motor neuron, the synaptic cleft, and the postsynaptic membrane of the muscle fiber. Upon the arrival of an action potential at the motor neuron terminal, the following sequence of events occurs:

- **Release of Acetylcholine**: The action potential causes voltage-gated calcium (Ca^{2+}) channels to open, leading to an influx of calcium ions. This influx triggers the fusion of synaptic vesicles containing ACh with the presynaptic membrane, resulting in the exocytosis of acetylcholine into the synaptic cleft.
- **Binding to Receptors**: Acetylcholine diffuses across the synaptic cleft and binds to nicotinic acetylcholine receptors (nAChRs) on the postsynaptic membrane of the muscle fiber. This binding induces a conformational change in the receptor, allowing sodium (Na^+) ions to enter the muscle cell and potassium (K^+) ions to exit.
- **Depolarization and Muscle Contraction**: The influx of sodium ions depolarizes the muscle cell membrane, generating an end-plate potential. If this depolarization reaches a certain threshold, it triggers an action potential in the muscle fiber. The action potential propagates along the muscle membrane and down the T-tubules, leading to the release of calcium ions from the sarcoplasmic reticulum. The released calcium ions interact with the contractile proteins (actin and myosin), resulting in muscle contraction.

Impact of ACh Inhibition on Muscle Performance

The regulation of acetylcholine levels at the NMJ is crucial for normal muscle function. Inhibition of acetylcholinesterase (AChE) can have significant effects on muscle performance.

1. Enhancement of Muscle Contraction

When AChE is inhibited, acetylcholine remains in the synaptic cleft for an extended period, leading to prolonged stimulation of nicotinic receptors. This prolonged exposure results in:

- **Increased Muscle Strength**: Enhanced levels of acetylcholine can lead to stronger and more sustained muscle contractions, which may be beneficial in certain medical conditions, such as myasthenia gravis, where muscle weakness is a primary symptom.
- **Potential for Overstimulation**: While increased ACh levels can enhance muscle contraction, excessive stimulation can lead to muscle fatigue and cramping due to continuous depolarization of the muscle membrane.

2. Cholinergic Crisis

In cases of excessive AChE inhibition, such as in poisoning scenarios or with the use of certain AChEIs at high doses, a condition known as a cholinergic crisis can occur. Symptoms include:

- **Severe Muscle Weakness**: Continuous stimulation leads to desensitization of nicotinic receptors, causing reduced responsiveness to further stimulation, resulting in profound muscle weakness.
- **Respiratory Failure**: In severe cases, the respiratory muscles can become paralyzed, leading to respiratory distress or failure, which is a medical emergency.

Clinical Applications

Understanding the role of acetylcholine in muscle function and the implications of AChE inhibition informs clinical practices, particularly in conditions characterized by muscle weakness or dysfunction.

1. Myasthenia Gravis

In myasthenia gravis, where the immune system attacks nicotinic receptors at the NMJ, the use of AChE inhibitors like pyridostigmine can significantly improve muscle strength by increasing acetylcholine levels. This therapeutic approach can enhance muscle performance and quality of life for patients.

2. Postoperative Recovery

AChE inhibitors may be considered in certain postoperative scenarios to improve muscle function and accelerate recovery from anesthesia-induced muscle weakness. Careful monitoring is essential to balance benefits against potential risks.

Summary

Acetylcholine is integral to muscle function, particularly at the neuromuscular junction, where it facilitates the communication between motor neurons and muscle fibers. The inhibition of acetylcholinesterase can enhance muscle performance through increased acetylcholine levels, leading to stronger and prolonged muscle contractions. However, excessive inhibition can lead to adverse effects, including cholinergic crisis. Understanding these dynamics is crucial for developing effective therapeutic strategies in neuromuscular disorders and optimizing muscle function.

Chapter 12: Acetylcholine and the Autonomic Nervous System

Role of Acetylcholine in the Sympathetic and Parasympathetic Systems

Acetylcholine (ACh) is a pivotal neurotransmitter in the autonomic nervous system (ANS), which regulates involuntary bodily functions such as heart rate, digestion, and respiratory rate. The ANS is divided into two main branches: the sympathetic and parasympathetic systems. Acetylcholine's role differs significantly between these two systems, influencing their distinct physiological effects.

1. Parasympathetic Nervous System

The parasympathetic nervous system (PNS) is often described as the "rest and digest" system. It promotes a state of relaxation and conservation of energy.

- **Acetylcholine Release**: In the PNS, acetylcholine is the primary neurotransmitter released by postganglionic neurons. When acetylcholine is released, it binds to muscarinic receptors on target organs, leading to various physiological responses.
- **Physiological Effects**: Activation of ACh in the PNS results in decreased heart rate, increased gastrointestinal activity, enhanced salivation, and contraction of the bladder. For example, ACh released onto the heart's pacemaker cells reduces heart rate by acting on muscarinic receptors (M2 subtype), leading to a calming effect on the body.

2. Sympathetic Nervous System

The sympathetic nervous system (SNS) prepares the body for "fight or flight" responses, increasing alertness and energy expenditure.

- **Acetylcholine's Role**: While the primary neurotransmitter of the sympathetic nervous system is norepinephrine, acetylcholine is still important in this system, particularly at the level of the ganglia and in certain effector tissues, such as sweat glands.
- **Physiological Effects**: In the SNS, acetylcholine is released from preganglionic neurons, where it binds to nicotinic receptors on postganglionic neurons. This action stimulates the release of norepinephrine from postganglionic sympathetic neurons, which then binds to adrenergic receptors on target tissues, resulting in increased heart rate, bronchodilation, and pupil dilation. Additionally, acetylcholine is directly involved in stimulating sweat glands, where it acts on muscarinic receptors.

Effects of AChE Inhibitors on Autonomic Regulation

The inhibition of acetylcholinesterase (AChE) can have profound effects on the autonomic nervous system by increasing the availability of acetylcholine at synapses. This has implications for both the sympathetic and parasympathetic branches.

1. Impact on the Parasympathetic System

Enhanced Parasympathetic Activity

2. Impact on the Sympathetic System

Altered Sympathetic Response

3. Clinical Considerations

- **Balance of Autonomic Tone**: The use of AChE inhibitors in clinical practice must be carefully monitored to avoid overstimulation of the autonomic nervous system. For instance, in patients with pre-existing conditions, such as asthma or cardiac issues, enhancing cholinergic activity could exacerbate symptoms.
- **Potential Therapeutic Benefits**: Despite the risks, AChE inhibitors hold potential therapeutic benefits for managing autonomic dysfunctions, including certain forms of orthostatic hypotension and gastrointestinal dysmotility. Research is ongoing to better understand the optimal use of these agents in various autonomic disorders.

Summary

Acetylcholine is a crucial neurotransmitter in the autonomic nervous system, playing distinct roles in both the sympathetic and parasympathetic systems. By enhancing cholinergic signaling through the inhibition of acetylcholinesterase, AChE inhibitors can significantly influence autonomic regulation, providing therapeutic benefits while necessitating careful monitoring to manage potential side effects. Understanding these dynamics is essential for optimizing treatment strategies that target autonomic dysfunctions and improve patient outcomes.

Chapter 13: Advances in AChE Inhibitor Research

Recent Discoveries and Innovations

The field of acetylcholinesterase inhibitor (AChEI) research has evolved significantly over recent years, driven by the urgent need for effective treatments for neurodegenerative diseases and other conditions associated with cholinergic dysfunction. This chapter explores the latest discoveries and innovations in AChEI research, highlighting novel compounds, mechanisms, and therapeutic applications.

1. New AChEI Compounds

Researchers have been actively developing new AChEIs with improved efficacy and safety profiles. Some recent advancements include:

- **Multitarget Drugs**: Recent studies have focused on designing multitarget drugs that not only inhibit AChE but also modulate other neurotransmitter systems or biological pathways involved in neuroprotection. For example, compounds that inhibit AChE and act as NMDA receptor antagonists are being investigated for their potential to enhance cognitive function while reducing excitotoxicity associated with neurodegeneration.
- **Natural Product Derivatives**: The exploration of natural compounds as AChEIs continues to gain traction. Recent studies have isolated and characterized novel AChEIs from various plant sources, showing promising results in preclinical models. These compounds often exhibit fewer side effects than synthetic AChEIs, making them attractive candidates for further development.
- **Nanotechnology**: Advances in nanotechnology have led to the development of nanoparticle-based AChEIs that can enhance drug delivery and efficacy. These nanoparticles can be designed to target specific tissues or cells, improving therapeutic outcomes while minimizing systemic side effects.

2. Mechanistic Insights

Recent research has also uncovered new insights into the mechanisms of AChE inhibition and its effects on cholinergic signaling:

- **Structural Biology**: Advances in structural biology, including X-ray crystallography and cryo-electron microscopy, have provided detailed insights into the structure of AChE and its interactions with various inhibitors. Understanding the molecular dynamics of AChE has enabled researchers to design more effective inhibitors that fit better within the enzyme's active site.
- **Allosteric Modulation**: Investigations into allosteric sites on AChE have revealed potential for developing allosteric modulators that enhance the enzyme's function without competing with acetylcholine. These compounds may provide a novel approach to regulating cholinergic signaling, particularly in conditions where traditional AChEIs may not be suitable.

Future Directions for AChE Inhibitor Development

As the understanding of acetylcholine's role in various physiological processes expands, so too does the potential for innovative AChEI therapies. Future research directions may include:

1. Personalized Medicine

With advancements in genetics and pharmacogenomics, there is a growing interest in personalized medicine approaches to AChEI therapy. Tailoring AChEI treatment based on individual genetic profiles and metabolic responses could optimize efficacy and minimize adverse effects.

2. Combination Therapies

Exploring combination therapies that integrate AChEIs with other pharmacological agents could enhance therapeutic outcomes. For instance, combining AChEIs with neuroprotective agents, antioxidants, or anti-inflammatory compounds may provide synergistic effects in treating neurodegenerative diseases.

3. Clinical Trials and Long-term Studies

Ongoing and future clinical trials will be crucial for evaluating the long-term safety and efficacy of new AChEIs and combination therapies. Longitudinal studies can provide insights into the progression of neurodegenerative diseases and the potential role of AChEIs in modifying disease trajectories.

4. Regulatory Challenges

As new AChEIs and therapeutic strategies emerge, navigating the regulatory landscape will be critical. Ensuring that new compounds meet safety and efficacy standards while being accessible to patients will require collaboration between researchers, regulatory agencies, and healthcare providers.

Summary

Advances in acetylcholinesterase inhibitor research are paving the way for innovative therapies to enhance cholinergic signaling and address conditions associated with cholinergic deficits. New compounds, mechanistic insights, and future directions emphasize the importance of continued research in this field. As the understanding of acetylcholine's role in health and disease deepens, AChEIs may play an increasingly critical role in developing effective treatments for neurodegenerative diseases and other cholinergic disorders.

Chapter 14: Acetylcholine and Mood Regulation

Connection Between Acetylcholine and Mental Health

Acetylcholine (ACh) is not only vital for cognitive functions such as learning and memory, but it also plays an essential role in regulating mood and emotional responses. Research has increasingly highlighted the connection between the cholinergic system and various mood disorders, providing insights into potential therapeutic interventions.

1. Acetylcholine's Role in Emotional Processing

Acetylcholine influences several brain regions involved in emotional regulation, particularly the hippocampus, amygdala, and prefrontal cortex:

- **Hippocampus**: ACh enhances synaptic plasticity in the hippocampus, which is crucial for forming emotional memories. Dysregulation of acetylcholine in this area may contribute to mood disorders, as emotional experiences are often intertwined with memory.
- **Amygdala**: The amygdala is responsible for processing emotions such as fear and pleasure. ACh modulates the activity of neurons in the amygdala, influencing emotional responses. A balanced cholinergic signaling in this region is important for appropriate emotional regulation.
- **Prefrontal Cortex**: This area is involved in decision-making, social behavior, and impulse control. Acetylcholine facilitates cognitive flexibility, allowing for adaptive emotional responses to changing situations. Alterations in cholinergic signaling here can affect mood stability and regulation.

Potential of AChE Inhibitors in Treating Mood Disorders

Given the role of acetylcholine in mood regulation, acetylcholinesterase inhibitors (AChEIs) have emerged as potential therapeutic agents for treating mood disorders. By increasing the availability of acetylcholine, AChEIs may help restore balance in the cholinergic system and alleviate symptoms of depression and anxiety.

1. Alzheimer's Disease and Depression

Patients with Alzheimer's disease often exhibit depressive symptoms alongside cognitive decline. Research indicates that AChEIs, commonly prescribed to enhance cognitive function, may also improve mood and overall quality of life. By augmenting cholinergic signaling, these drugs can potentially alleviate depressive symptoms in this population.

2. AChE Inhibition in Other Mood Disorders

AChEIs are being investigated for their effects on other mood disorders, such as major depressive disorder (MDD) and generalized anxiety disorder (GAD). Preliminary studies suggest that enhancing cholinergic activity may provide therapeutic benefits for individuals experiencing these conditions. AChEIs could work synergistically with traditional antidepressants, improving treatment outcomes.

Research and Future Directions

Ongoing research aims to further elucidate the relationship between acetylcholine and mood regulation, with a focus on several key areas:

1. Neurochemical Mechanisms

Understanding the neurochemical pathways through which ACh influences mood will provide insights into developing targeted therapies. Investigating how ACh interacts with other neurotransmitter systems, such as serotonin and dopamine, can help clarify its role in emotional regulation.

2. Genetic Variations

Research into genetic variations affecting acetylcholine metabolism may reveal individualized responses to AChEIs. Identifying biomarkers associated with cholinergic signaling could lead to personalized treatment approaches for mood disorders.

3. Clinical Trials

Clinical trials exploring the efficacy of AChEIs in treating mood disorders are crucial. Rigorous studies assessing their impact on depressive symptoms and anxiety levels will help establish AChEIs as viable options in the psychopharmacological arsenal.

Summary

Acetylcholine is intricately linked to mood regulation, influencing emotional processing and stability through its actions in key brain regions. The potential of acetylcholinesterase inhibitors to alleviate symptoms of mood disorders presents an exciting avenue for research and therapeutic development. As our understanding of the cholinergic system and its effects on mental health grows, AChEIs may play a vital role in treating mood-related conditions, offering hope for improved outcomes in affected individuals.

Chapter 15: Acetylcholine in Pain Modulation

Role of Acetylcholine in Pain Pathways

Acetylcholine (ACh) is primarily recognized for its role as a neurotransmitter in the central and peripheral nervous systems. However, its involvement in pain modulation is gaining increasing attention in both research and clinical contexts. ACh plays a multifaceted role in pain pathways, influencing the perception and transmission of pain signals through various mechanisms.

1. Cholinergic Pathways in Pain Processing

ACh contributes to the modulation of pain at multiple levels, including peripheral nociceptive pathways and central nervous system processes. Key components include:

- **Peripheral Mechanisms**: Acetylcholine is involved in the activation of sensory neurons that respond to noxious stimuli. Cholinergic receptors located on nociceptors can enhance the sensitivity of these neurons, influencing how pain is perceived. The interplay between ACh and other neurotransmitters at the site of injury can amplify pain signals, contributing to the sensation of pain.

- **Central Mechanisms**: Within the spinal cord and brain, ACh influences pain modulation by interacting with various receptors and signaling pathways. Cholinergic projections from the brainstem to the spinal cord play a role in descending pain modulation, where they can either facilitate or inhibit pain transmission. This dual role can depend on the type of cholinergic receptor activated (e.g., nicotinic vs. muscarinic receptors).

AChE Inhibition and Pain Management Strategies

Given the role of ACh in pain pathways, the inhibition of acetylcholinesterase (AChE) presents a potential therapeutic strategy for managing pain. By preventing the breakdown of ACh, AChE inhibitors can enhance cholinergic signaling, thereby influencing pain perception and modulation.

1. Potential Benefits of AChE Inhibitors in Pain Relief

- **Enhanced Analgesic Effects**: AChE inhibitors may enhance the analgesic effects of certain pain medications by promoting cholinergic activity. For instance, studies suggest that combining AChEIs with opioids may improve pain relief and reduce the required dosage of opioids, potentially minimizing their side effects and the risk of dependence.
- **Neuropathic Pain Management**: Neuropathic pain, resulting from nerve damage or dysfunction, may also be alleviated through AChE inhibition. By enhancing cholinergic signaling, AChE inhibitors may help restore balance in the pain pathways, reducing the severity of neuropathic pain symptoms.

2. Clinical Applications and Research

While the potential for AChE inhibitors in pain management is promising, clinical applications are still being explored. Current research focuses on:

- **Pain Syndromes**: Investigating the efficacy of AChE inhibitors in various pain syndromes, including fibromyalgia, diabetic neuropathy, and post-surgical pain, is crucial for understanding their therapeutic potential.
- **Mechanistic Studies**: Ongoing studies are examining the mechanisms by which ACh and AChE inhibitors modulate pain pathways. This includes exploring the interactions between ACh, other neurotransmitters, and inflammatory mediators.

Challenges and Considerations

Despite the potential benefits, there are challenges associated with using AChE inhibitors for pain management:

- **Side Effects**: The systemic effects of AChE inhibition can lead to unwanted side effects, including increased gastrointestinal motility, muscle cramps, and bradycardia. Careful monitoring and dose adjustments are necessary to mitigate these risks.
- **Complexity of Pain Mechanisms**: Pain is a complex and multifactorial experience, influenced by various biological, psychological, and social factors. AChE inhibitors may not be effective for all types of pain, and their use should be tailored to individual patient needs.

Summary

Acetylcholine plays a significant role in pain modulation through its actions in both peripheral and central pain pathways. The inhibition of acetylcholinesterase presents a potential strategy for enhancing pain management, particularly in conditions like neuropathic pain and as an adjunct to other analgesics. Ongoing research into the mechanisms of ACh and the clinical applications of AChE inhibitors will continue to shed light on their role in pain management, providing new avenues for therapeutic intervention in chronic pain conditions.

Chapter 16: Genetic Variations and Acetylcholine Function

Influence of Genetics on Acetylcholine Metabolism

The functionality and effectiveness of the cholinergic system, driven by acetylcholine (ACh), are significantly influenced by genetic variations. These genetic factors can affect everything from the synthesis, release, and degradation of acetylcholine to the sensitivity of acetylcholine receptors. Understanding these variations is crucial for personalized medicine, especially in the context of treating disorders related to cholinergic dysfunction.

1. Genetic Polymorphisms Affecting AChE Activity

Acetylcholinesterase (AChE) is the enzyme responsible for breaking down acetylcholine in the synaptic cleft. Variations in the gene encoding AChE can lead to differences in enzyme activity, influencing how quickly acetylcholine is cleared from the synapse. Key polymorphisms include:

- **AChE Gene Variants**: Certain single nucleotide polymorphisms (SNPs) in the AChE gene can lead to altered enzyme activity. For example, variations may result in faster or slower degradation of acetylcholine, impacting cholinergic signaling and the overall cholinergic tone in the nervous system.
- **Impact on Drug Response**: Individuals with different AChE polymorphisms may respond variably to AChE inhibitors, such as those used in Alzheimer's disease. Genetic testing may help predict which patients are likely to benefit most from AChEIs based on their metabolic profile.

2. Cholinergic Receptor Variants

In addition to AChE, genetic variations also affect the nicotinic and muscarinic acetylcholine receptors. Variants in the genes encoding these receptors can lead to differences in receptor density, binding affinity, and signaling pathways.

- **Nicotinic Receptor Variants**: Genetic polymorphisms in the CHRNA5, CHRNA3, and CHRNB4 genes, which encode subunits of nicotinic receptors, have been associated with altered responses to nicotine and differences in pain perception, addiction susceptibility, and cognitive function. Variants can influence how these receptors respond to acetylcholine and other agonists.
- **Muscarinic Receptor Variants**: Variations in muscarinic receptor genes (e.g., CHRM1, CHRM2) can also impact cholinergic signaling and influence cognitive processes, mood regulation, and response to various medications targeting the cholinergic system.

Implications for Personalized Medicine

Understanding genetic variations in the cholinergic system has significant implications for personalized medicine, particularly in the treatment of neurological and psychiatric disorders.

1. Tailored Therapeutic Approaches

Genetic testing may help identify individuals who are more likely to benefit from AChE inhibitors or other cholinergic agents. By understanding an individual's genetic makeup, clinicians can tailor treatments to enhance efficacy and minimize side effects.

- **Optimizing Dosages**: Genetic variations affecting AChE activity could guide the optimal dosing of AChE inhibitors, ensuring patients receive the most effective dose with the least risk of adverse effects.
- **Combination Therapies**: Personalized approaches may involve combining AChE inhibitors with other medications that target different pathways in patients with specific genetic profiles, potentially leading to better overall treatment outcomes.

2. Research Directions

Ongoing research aims to better understand the relationship between genetics and cholinergic function. Areas of exploration include:

- **Large-scale Genetic Studies**: Genome-wide association studies (GWAS) can help identify novel genetic variants associated with cholinergic dysfunction and their effects on disease susceptibility and treatment response.
- **Functional Genomics**: Investigating how specific genetic variants affect the expression and function of cholinergic components can shed light on mechanisms underlying individual differences in cholinergic signaling and associated disorders.

Summary

Genetic variations significantly influence acetylcholine metabolism and cholinergic signaling, affecting individual responses to therapies targeting the cholinergic system. Understanding these genetic factors is crucial for advancing personalized medicine, optimizing treatment strategies, and improving outcomes for patients with disorders related to cholinergic dysfunction. As research continues to uncover the complexities of genetic influences on acetylcholine function, the potential for tailored therapeutic approaches in neurological and psychiatric care will grow.

Chapter 17: Acetylcholine and Neurodegeneration

Relationship Between Acetylcholine Deficits and Neurodegenerative Diseases

Acetylcholine (ACh) is a neurotransmitter that plays a critical role in various cognitive and physiological functions. Its deficits have been implicated in several neurodegenerative diseases, most notably Alzheimer's disease, but also including other conditions such as Parkinson's disease and Huntington's disease. Understanding the relationship between acetylcholine deficits and neurodegeneration is essential for developing effective therapeutic strategies.

1. Alzheimer's Disease

Alzheimer's disease (AD) is characterized by progressive cognitive decline, memory loss, and behavioral changes. One of the hallmark features of AD is the significant loss of cholinergic neurons in the basal forebrain, leading to decreased levels of acetylcholine in the cerebral cortex and hippocampus.

- **Pathophysiology**: In AD, neurofibrillary tangles and amyloid plaques disrupt cholinergic signaling pathways. The loss of ACh contributes to impairments in synaptic plasticity and cognitive function, particularly in learning and memory.
- **AChE Inhibitors**: Acetylcholinesterase inhibitors (AChEIs), such as donepezil, rivastigmine, and galantamine, are commonly used in clinical practice to treat Alzheimer's disease. These drugs work by increasing acetylcholine levels in the brain, thereby enhancing cholinergic transmission and improving cognitive symptoms in some patients.

2. Parkinson's Disease

While Parkinson's disease (PD) is primarily characterized by dopaminergic neuron loss, acetylcholine also plays a significant role in its pathology.

- **Cholinergic Imbalance**: In PD, there is an imbalance between cholinergic and dopaminergic systems. The loss of dopamine in the basal ganglia can lead to increased cholinergic activity, contributing to symptoms such as tremors, rigidity, and bradykinesia. This imbalance highlights the complex interactions between neurotransmitter systems in neurodegenerative diseases.
- **Potential Therapeutic Approaches**: Research is ongoing into the use of AChE inhibitors as adjunctive therapies in Parkinson's disease. By restoring some cholinergic activity, these drugs may help alleviate certain symptoms associated with dopaminergic deficits.

3. Huntington's Disease

Huntington's disease (HD) is a hereditary neurodegenerative disorder characterized by motor dysfunction, cognitive decline, and psychiatric symptoms.

- **Cognitive Impairment**: Individuals with HD exhibit significant cognitive decline, which is linked to cholinergic dysfunction. Research suggests that cholinergic deficits in the striatum and other regions contribute to the cognitive and behavioral symptoms of the disease.
- **Cholinergic Therapies**: Exploring cholinergic therapies, including AChE inhibitors, may offer potential benefits for managing cognitive symptoms in HD, although more research is needed to establish efficacy.

Role of AChE Inhibitors in Disease Progression

The use of acetylcholinesterase inhibitors in neurodegenerative diseases extends beyond symptom management; they may also have implications for disease progression and neuroprotection.

1. Neuroprotective Effects

Recent studies suggest that AChEIs may exert neuroprotective effects by promoting cholinergic signaling and supporting synaptic health. Increased ACh levels can enhance neuronal resilience against neurotoxic factors associated with neurodegeneration, potentially slowing disease progression.

Synaptic Plasticity

2. Combining AChEIs with Other Therapies

Ongoing research is investigating the potential benefits of combining AChEIs with other neuroprotective agents, such as antioxidants, anti-inflammatory drugs, and drugs targeting other neurotransmitter systems. This multi-targeted approach may enhance overall therapeutic efficacy and improve patient outcomes.

Challenges and Considerations

Despite the potential benefits of AChE inhibitors, challenges remain in their application for neurodegenerative diseases:

- **Patient Variability**: Individual responses to AChE inhibitors can vary significantly based on genetic factors, disease stage, and co-existing conditions. Personalized medicine approaches, including pharmacogenomic testing, may help tailor treatments to optimize efficacy and minimize side effects.
- **Long-term Efficacy**: While AChEIs can provide symptomatic relief, their long-term efficacy in modifying disease progression remains a topic of debate. Continued research is necessary to clarify their role in the context of neurodegeneration.

Summary

Acetylcholine deficits are intricately linked to the pathology of several neurodegenerative diseases, including Alzheimer's, Parkinson's, and Huntington's diseases. Understanding these relationships is crucial for developing effective therapeutic strategies, particularly with the use of acetylcholinesterase inhibitors. As research progresses, the potential for AChEIs to not only alleviate symptoms but also exert neuroprotective effects offers hope for improved outcomes in patients with neurodegenerative conditions.

Chapter 18: Dietary Influences on Acetylcholine Levels

Nutritional Factors Affecting Acetylcholine Synthesis

Acetylcholine (ACh) synthesis and availability in the body can be influenced significantly by dietary choices. Certain nutrients play essential roles in the production of acetylcholine, and understanding these relationships can help in the management of conditions associated with cholinergic dysfunction. This chapter explores the dietary influences on acetylcholine levels, highlighting key nutrients, food sources, and potential dietary strategies.

1. Key Nutrients for Acetylcholine Production

The synthesis of acetylcholine is a complex process that relies on several critical nutrients:

Choline

Food Sources

Folic Acid (Vitamin B9)

Food Sources

Vitamin B12

Food Sources

Omega-3 Fatty Acids

Food Sources

Foods and Supplements That Support Acetylcholine Production

Incorporating specific foods and supplements into the diet can help enhance acetylcholine levels and support overall cognitive function.

1. Dietary Approaches

- **Balanced Diet**: A balanced diet that includes sufficient protein, healthy fats, and a variety of fruits and vegetables can ensure an adequate supply of the nutrients needed for acetylcholine synthesis.
- **High-Choline Foods**: Including high-choline foods, such as eggs and liver, in the diet can directly increase acetylcholine production.
- **Supplementation**: For individuals who may not get enough choline from their diet, supplementation may be beneficial. Choline bitartrate and alpha-GPC (alpha-glycerylphosphorylcholine) are popular choline supplements that can enhance choline levels in the body.

2. Potential Impact of Diet on Cognitive Function

Emerging research suggests that diets rich in choline and other supportive nutrients may help improve cognitive function and reduce the risk of cognitive decline:

- **Mediterranean Diet**: The Mediterranean diet, characterized by high consumption of fruits, vegetables, whole grains, fish, nuts, and healthy fats, has been associated with cognitive benefits. This diet provides essential nutrients that support acetylcholine synthesis and overall brain health.
- **Cognitive Health**: Studies have shown that adequate choline intake is linked to better cognitive performance and a lower risk of neurodegenerative diseases, suggesting that dietary strategies may play a preventive role in cognitive decline.

Conclusion

Diet plays a critical role in influencing acetylcholine levels and, consequently, cholinergic signaling. Key nutrients such as choline, folic acid, vitamin B12, and omega-3 fatty acids are essential for the synthesis and function of acetylcholine. Incorporating a balanced diet rich in these nutrients can support optimal cholinergic function and potentially mitigate the risk of cognitive decline. As research continues to explore the connections between diet, acetylcholine, and brain health, dietary strategies may emerge as valuable tools in managing conditions associated with cholinergic dysfunction.

Chapter 19: Exploring AChE Inhibitors in Sports and Performance

Use of AChE Inhibitors in Enhancing Athletic Performance

The relationship between acetylcholine (ACh) and physical performance is complex and multifaceted. Acetylcholine plays a crucial role in neuromuscular transmission, impacting muscle contraction and overall physical endurance. This chapter explores the potential use of acetylcholinesterase inhibitors (AChEIs) in sports and athletic performance, considering both the mechanisms by which they may enhance performance and the ethical implications of their use.

1. Mechanisms of Action in Muscle Function

ACh is essential for the communication between motor neurons and muscle fibers at the neuromuscular junction. When a motor neuron releases acetylcholine, it binds to nicotinic receptors on the muscle membrane, leading to muscle contraction. AChE breaks down acetylcholine in the synaptic cleft, terminating the signal. By inhibiting AChE, AChEIs can prolong the action of acetylcholine, potentially leading to:

- **Increased Muscle Contraction**: With prolonged availability of acetylcholine, muscles may experience enhanced contractile force and endurance. This could be particularly beneficial in high-intensity or prolonged physical activities.
- **Improved Reaction Time**: Enhanced cholinergic signaling may improve neuromuscular transmission speed, leading to quicker muscle response times. This could be advantageous in sports requiring rapid movements and reflexes.
- **Potential for Greater Endurance**: AChEIs might help reduce fatigue during prolonged exercise by maintaining effective neuromuscular communication and delaying the onset of muscle fatigue.

2. Research and Evidence

While the theoretical benefits of AChEIs in sports are compelling, empirical research is limited. Studies exploring the effects of AChE inhibition on athletic performance are still emerging, with early findings suggesting potential advantages, particularly in resistance training and endurance sports.

- **Pilot Studies**: Some pilot studies have indicated improvements in strength and endurance among athletes using AChEIs compared to placebo groups. However, these studies often involve small sample sizes and require further validation through larger, more rigorous trials.
- **Cognitive and Motor Coordination**: Research has also suggested that AChEIs may improve cognitive function and motor coordination, critical components of athletic performance. Enhanced focus, decision-making, and reaction time can positively influence an athlete's overall performance.

Ethical Considerations and Regulations in Sports

The use of AChE inhibitors in sports raises significant ethical questions, particularly regarding fairness and safety. Various sports organizations, including the World Anti-Doping Agency (WADA), impose strict regulations on the use of performance-enhancing substances, which may include AChEIs.

1. Doping Regulations

- **Classifications**: Depending on their mechanism of action and potential effects on performance, AChE inhibitors could be classified as performance-enhancing drugs. Athletes must be cautious about using these substances without thorough understanding and legal clearance.
- **Testing and Compliance**: Athletes competing at elite levels are subject to rigorous drug testing, and the inclusion of AChEIs in the prohibited substances list would have significant implications for compliance and potential penalties for athletes.

2. Health Risks

The safety profile of AChE inhibitors must be carefully considered in the context of athletic performance. Possible side effects, including muscle cramps, gastrointestinal disturbances, and cardiovascular effects, could pose risks to athletes, particularly in high-stress environments such as competitive sports.

Long-term Effects

Summary

Acetylcholine plays a vital role in neuromuscular function and athletic performance, and the potential use of acetylcholinesterase inhibitors to enhance physical capabilities presents an intriguing area of study. While preliminary research suggests possible benefits, the ethical implications and health risks associated with AChEIs in sports warrant careful consideration. As the understanding of cholinergic signaling continues to evolve, future research will be essential to determine the appropriate role of AChE inhibitors in enhancing athletic performance while ensuring fair and safe practices in sports.

Chapter 20: Acetylcholine's Role in the Endocrine System

Interaction Between Acetylcholine and Hormone Regulation

Acetylcholine (ACh), primarily recognized for its role as a neurotransmitter in the central and peripheral nervous systems, also exerts significant influence on the endocrine system. Understanding the interactions between acetylcholine and hormonal regulation reveals the complexity of physiological processes and highlights the potential therapeutic applications of targeting the cholinergic system in endocrine disorders.

1. Cholinergic Signaling in Endocrine Glands

Acetylcholine acts on various endocrine glands, modulating hormone secretion and influencing physiological functions. Key interactions include:

- **Pancreas**: In the pancreas, acetylcholine stimulates the release of insulin from pancreatic β-cells. This cholinergic stimulation is particularly significant during the postprandial state when glucose levels rise. The binding of ACh to muscarinic receptors on β-cells promotes insulin secretion, helping to regulate blood glucose levels.
- **Adrenal Glands**: The adrenal medulla responds to cholinergic signaling by releasing catecholamines, such as adrenaline and noradrenaline, during stress responses. ACh binds to nicotinic receptors on chromaffin cells in the adrenal medulla, leading to the secretion of these hormones, which prepare the body for "fight or flight" responses.
- **Thyroid Gland**: Acetylcholine also influences thyroid function. Cholinergic signaling can stimulate thyroid hormone secretion, which plays a crucial role in metabolism, growth, and development. The interaction between ACh and the thyroid gland may contribute to the regulation of metabolic processes.

Effects of AChE Inhibition on Endocrine Functions

Inhibiting acetylcholinesterase (AChE), the enzyme responsible for breaking down acetylcholine, can enhance cholinergic signaling and potentially affect endocrine functions. The implications of AChE inhibition for hormonal regulation are significant:

1. Enhanced Hormonal Secretion

Increased availability of acetylcholine due to AChE inhibition may enhance the secretion of various hormones:

- **Insulin Release**: As ACh promotes insulin secretion from pancreatic β-cells, AChE inhibitors may increase insulin release, which could be beneficial for managing blood glucose levels in conditions like type 2 diabetes. This effect requires careful monitoring, as excessive insulin release could lead to hypoglycemia.
- **Stress Response**: In situations of stress or anxiety, AChE inhibitors might enhance the adrenal medulla's response, increasing the secretion of catecholamines. While this can be beneficial in acute stress situations, chronic overactivation may lead to adverse effects, including hypertension and anxiety disorders.

2. Impact on Metabolism

Cholinergic modulation of endocrine function can influence metabolic processes:

- **Thyroid Function**: Enhanced cholinergic signaling through AChE inhibition may lead to increased thyroid hormone secretion, affecting metabolic rate and energy expenditure. This effect could have therapeutic implications for conditions characterized by metabolic dysregulation.
- **Overall Energy Homeostasis**: ACh plays a role in appetite regulation and energy balance. By enhancing cholinergic signaling, AChE inhibitors could impact weight management and metabolic health, but more research is needed to understand these interactions fully.

Clinical Implications and Future Directions

The interaction between acetylcholine and the endocrine system opens avenues for novel therapeutic approaches in managing endocrine disorders:

1. Potential for Treatment of Diabetes and Metabolic Disorders

AChE inhibitors may offer new strategies for treating type 2 diabetes and other metabolic disorders by improving insulin secretion and glucose homeostasis. Further research is warranted to explore the safety and efficacy of these compounds in clinical settings.

2. Understanding Endocrine Disorders

Investigating the role of cholinergic signaling in endocrine disorders may lead to better diagnostic and treatment options. Conditions such as adrenal insufficiency, thyroid disorders, and metabolic syndrome could benefit from a deeper understanding of ACh's role.

Summary

Acetylcholine plays a crucial role in the regulation of the endocrine system, influencing hormone secretion and metabolic processes. The inhibition of acetylcholinesterase can enhance cholinergic signaling, with potential implications for managing endocrine disorders, particularly diabetes and metabolic dysfunctions. Continued research into the interplay between the cholinergic system and endocrine function will be vital for developing effective therapeutic strategies and improving patient outcomes in conditions influenced by these complex interactions.

Chapter 21: The Future of Acetylcholine Research

Emerging Trends and Potential Breakthroughs

As the understanding of acetylcholine (ACh) and its multifaceted roles in the nervous system continues to evolve, research into its therapeutic applications and mechanisms is poised for significant advancements. This chapter explores emerging trends and potential breakthroughs in the field of acetylcholine research, highlighting the importance of interdisciplinary approaches and innovative methodologies.

1. Novel Therapeutic Targets

Recent research has identified new targets within the cholinergic system that could lead to the development of innovative therapies for a range of conditions:

- **Allosteric Modulators**: The development of allosteric modulators of acetylcholine receptors offers a promising avenue for enhancing cholinergic signaling. These compounds can selectively modulate receptor activity without fully activating the receptor, potentially providing therapeutic benefits with fewer side effects compared to traditional agonists.
- **Dual-Action Drugs**: Research into compounds that can simultaneously target multiple neurotransmitter systems, including acetylcholine, offers the potential for more effective treatments. For instance, drugs that act on both cholinergic and serotonergic systems may provide new strategies for treating mood disorders and cognitive impairments.

2. Advancements in Drug Delivery Systems

The efficacy of acetylcholine-based therapies may be enhanced through novel drug delivery systems that ensure targeted action and minimize systemic side effects:

- **Nanotechnology**: The use of nanoparticles and nanocarriers to deliver AChE inhibitors or other cholinergic agents directly to specific brain regions could enhance therapeutic outcomes while reducing side effects.
- **Intranasal Administration**: Emerging research on intranasal delivery methods for cholinergic drugs may provide rapid and effective access to the central nervous system, improving treatment efficacy for conditions such as Alzheimer's disease.

3. Personalized Medicine Approaches

With advances in genomics and pharmacogenomics, the future of acetylcholine research is likely to include personalized medicine approaches that tailor treatments to individual patient profiles:

- **Genetic Profiling**: Understanding an individual's genetic variations related to acetylcholine metabolism, receptor sensitivity, and overall cholinergic function can guide clinicians in selecting the most appropriate therapeutic strategies.
- **Biomarker Development**: The identification of biomarkers related to cholinergic dysfunction may facilitate early diagnosis and monitoring of treatment responses, enabling more effective management of neurodegenerative diseases and other cholinergic disorders.

Interdisciplinary Approaches to Studying Acetylcholine

The complexity of acetylcholine's role in various physiological and pathological processes necessitates interdisciplinary research approaches that integrate insights from multiple fields:

1. Neuroscience and Pharmacology

Collaborative efforts between neuroscientists and pharmacologists are crucial for developing a comprehensive understanding of cholinergic signaling. Research focused on the molecular mechanisms of ACh signaling, receptor interactions, and the pharmacodynamics of AChE inhibitors can lead to more effective therapies.

2. Nutrition and Lifestyle Research

Investigating the impact of dietary factors and lifestyle choices on acetylcholine levels and cholinergic function is an area ripe for exploration. Integrating nutritional science with pharmacology can help identify dietary strategies that support cholinergic health and enhance the efficacy of pharmacological interventions.

3. Psychology and Behavioral Sciences

Understanding the relationship between acetylcholine and cognitive functions, mood regulation, and behavior will benefit from interdisciplinary collaboration with psychology and behavioral sciences. Research in these areas can help elucidate the broader implications of cholinergic signaling in mental health and cognitive performance.

Conclusion

The future of acetylcholine research holds immense potential for uncovering novel therapeutic targets, improving drug delivery systems, and advancing personalized medicine approaches. Interdisciplinary collaboration will be essential in driving innovative research that enhances our understanding of acetylcholine's complex role in health and disease. As research continues to evolve, the potential for breakthroughs in treating neurodegenerative diseases, mood disorders, and other cholinergic dysfunctions will expand, offering hope for improved patient outcomes and enhanced quality of life.

Chapter 22: Case Studies: Successful AChE Inhibitor Treatments

Introduction

The therapeutic potential of acetylcholinesterase inhibitors (AChEIs) has been explored across various clinical settings, particularly in the treatment of neurodegenerative diseases such as Alzheimer's disease and myasthenia gravis. This chapter presents a selection of case studies that exemplify the efficacy of AChEIs in improving patient outcomes, alongside insights into the lessons learned from these applications.

Case Study 1: AChE Inhibitors in Alzheimer's Disease

Patient Profile

- **Patient**: Mr. Smith
- **Age**: 72
- **Diagnosis**: Mild to moderate Alzheimer's disease
- **Medications**: Initiated on Donepezil (an AChEI)

Treatment Overview

Mr. Smith was diagnosed with Alzheimer's disease after experiencing cognitive decline characterized by memory loss and difficulty in daily activities. His physician prescribed donepezil, aiming to improve his cognitive function by increasing acetylcholine levels in the brain.

Outcomes

- **Cognitive Improvement**: Over a six-month treatment period, assessments indicated a modest improvement in cognitive function, particularly in memory recall and daily living activities as measured by the Alzheimer's Disease Assessment Scale-Cognitive Subscale (ADAS-Cog).
- **Quality of Life**: Family members reported enhanced engagement and social interaction, contributing to a better overall quality of life.

Lessons Learned

- **Early Intervention**: Initiating AChEIs at an early stage of Alzheimer's disease can optimize cognitive function and delay progression.
- **Importance of Support**: The role of family support and education in managing the patient's condition was vital to sustaining improvements in daily functioning.

Case Study 2: AChE Inhibitors in Myasthenia Gravis

Patient Profile

- **Patient**: Ms. Johnson
- **Age**: 55
- **Diagnosis**: Myasthenia gravis
- **Medications**: Treated with Pyridostigmine, an AChEI

Treatment Overview

Ms. Johnson presented with muscle weakness and fatigue, characteristic of myasthenia gravis. Pyridostigmine was introduced to enhance neuromuscular transmission by inhibiting acetylcholinesterase and prolonging the action of acetylcholine at the neuromuscular junction.

Outcomes

- **Muscle Strength**: Following treatment, Ms. Johnson reported significant improvements in muscle strength and endurance, particularly in the upper limbs and facial muscles.
- **Functional Independence**: The patient regained the ability to perform daily activities independently, such as grooming and meal preparation.

Lessons Learned

- **Dosing and Timing**: Tailoring the dosing schedule of pyridostigmine to align with the patient's daily activities maximized therapeutic benefits.
- **Monitoring for Side Effects**: Regular monitoring for side effects, such as excessive salivation and gastrointestinal discomfort, was crucial to maintaining adherence to treatment.

Case Study 3: AChE Inhibitors in Cognitive Enhancement
Patient Profile

- **Patient**: Mr. Lee
- **Age**: 30
- **Diagnosis**: Attention Deficit Hyperactivity Disorder (ADHD)
- **Medications**: Off-label use of Donepezil

Treatment Overview

Mr. Lee, diagnosed with ADHD, was given donepezil to assess its effects on cognitive enhancement, specifically focusing on attention and memory. This off-label use was predicated on emerging research suggesting potential benefits in cognitive performance.

Outcomes

- **Improved Attention**: The patient reported increased focus and sustained attention during tasks, alongside subjective improvements in memory retention.
- **Enhanced Academic Performance**: Following the intervention, Mr. Lee's academic performance showed marked improvement, with better grades and engagement in class.

Lessons Learned

- **Caution with Off-Label Use**: While promising results were noted, further research is necessary to establish safety and efficacy for off-label use of AChEIs in non-neurodegenerative conditions.
- **Need for Monitoring**: Close monitoring of cognitive function and side effects is essential when utilizing AChEIs in populations outside traditional indications.

Conclusion

These case studies illustrate the potential of acetylcholinesterase inhibitors in treating various conditions, highlighting their effectiveness in improving cognitive function and muscle strength. The lessons learned from these cases emphasize the importance of early intervention, personalized treatment plans, and careful monitoring of side effects. As research continues to evolve, the role of AChEIs may expand, offering new avenues for enhancing patient outcomes in diverse clinical contexts.

Through ongoing clinical evaluation and interdisciplinary collaboration, the understanding and application of AChEIs will continue to advance, ultimately benefiting a broader range of patients facing cholinergic dysfunctions.

Chapter 23: Practical Applications: Using AChE Inhibitors Wisely

Introduction

Acetylcholinesterase inhibitors (AChEIs) have emerged as critical therapeutic agents in managing various neurological disorders and conditions associated with cholinergic dysfunction. However, their use must be approached with caution to maximize benefits while minimizing risks. This chapter outlines practical applications of AChEIs, providing guidelines for practitioners and patients on how to use these medications wisely and effectively.

1. Understanding the Indications for AChEIs

Before prescribing AChEIs, it is essential to understand their appropriate indications:

- **Alzheimer's Disease**: AChEIs like donepezil, rivastigmine, and galantamine are commonly used to treat mild to moderate Alzheimer's disease. They can help improve cognitive function and slow disease progression.
- **Myasthenia Gravis**: Pyridostigmine is an AChEI used to enhance neuromuscular transmission and improve muscle strength in patients with myasthenia gravis.
- **Other Neurological Conditions**: AChEIs may also be beneficial in conditions like Lewy body dementia and certain types of cognitive impairment.

2. Patient Assessment and Monitoring

A thorough assessment of the patient's medical history and current health status is crucial before initiating treatment with AChEIs. Key considerations include:

- **Medical History**: Evaluate for any contraindications, such as hypersensitivity to AChEIs, asthma, or other pulmonary conditions that could be exacerbated.
- **Concurrent Medications**: Review the patient's current medication regimen to avoid potential drug interactions. Some medications, particularly anticholinergic drugs, may counteract the effects of AChEIs.
- **Baseline Assessments**: Conduct baseline cognitive assessments, muscle strength tests, and vital sign measurements. These will serve as references to monitor treatment efficacy and side effects.

3. Dosing Strategies

Determining the appropriate dose of AChEIs is critical for optimizing therapeutic effects while minimizing side effects:

- **Start Low, Go Slow**: Begin with a low dose and titrate upward as tolerated. This approach helps mitigate potential side effects, such as gastrointestinal discomfort or increased muscle cramps.
- **Individualized Dosing**: Tailor dosing based on patient response and tolerance. Regular follow-up assessments are necessary to determine if dose adjustments are needed.

4. Educating Patients and Caregivers

Patient education is vital for the successful use of AChEIs:

- **Understanding Treatment Goals**: Clearly explain the purpose of AChEIs, including expected benefits and potential side effects. Educating patients about the chronic nature of conditions like Alzheimer's disease can help manage expectations.
- **Monitoring Side Effects**: Instruct patients and caregivers to monitor for common side effects, including nausea, diarrhea, fatigue, and muscle cramps. Encourage them to report any severe or persistent side effects promptly.
- **Lifestyle Considerations**: Discuss the importance of a healthy lifestyle, including regular exercise, a balanced diet, and cognitive engagement, to enhance the effects of AChEIs.

5. Monitoring Efficacy and Adjusting Treatment

Regular follow-up appointments are essential to assess the efficacy of AChEIs and make necessary adjustments:

- **Cognitive and Functional Assessments**: Utilize standardized assessment tools to measure changes in cognitive function and daily living activities over time.
- **Evaluating Side Effects**: Continuously evaluate for side effects and their impact on the patient's quality of life. If side effects are significant, consider dose adjustments or alternative treatments.
- **Long-Term Management**: Recognize that AChEIs may not halt disease progression but can provide symptomatic relief. Encourage ongoing discussions about long-term management strategies.

6. Considerations for Special Populations

Certain populations may require special consideration when using AChEIs:

- **Elderly Patients**: Older adults may be more sensitive to the side effects of AChEIs. Careful monitoring and gradual dosing adjustments are essential.
- **Patients with Comorbid Conditions**: Consider comorbidities such as cardiovascular disease, pulmonary issues, or gastrointestinal disorders that may influence treatment decisions.
- **Pediatric and Adolescent Patients**: While AChEIs are primarily used in adults, some studies suggest potential benefits in specific pediatric populations. Further research is needed in this area.

Conclusion

The practical application of acetylcholinesterase inhibitors in clinical settings offers promising benefits for patients with various neurological disorders. However, the safe and effective use of these agents requires careful assessment, patient education, and ongoing monitoring. By following the outlined guidelines, practitioners can optimize treatment outcomes and enhance the quality of life for their patients while minimizing risks associated with AChEIs.

Chapter 24: Conclusion: Mastering Acetylcholine

Summary of Key Concepts

Throughout this book, we have explored the multifaceted role of acetylcholine (ACh) in the nervous system, its implications for various physiological functions, and the therapeutic potential of acetylcholinesterase inhibitors (AChEIs). Key concepts include:

1. **The Role of Acetylcholine**: ACh is a crucial neurotransmitter that facilitates communication between neurons and muscles, playing a significant role in neurotransmission, muscle contraction, and cognitive functions such as learning and memory.

2. **Acetylcholine Receptor System**: The two main types of receptors, nicotinic and muscarinic, have distinct mechanisms of action and physiological effects. Understanding these receptors is vital for developing targeted therapies.

3. **Synthesis and Release**: The biosynthesis of ACh and its release at the neuromuscular junction are critical processes that underpin effective communication in the nervous system.

4. **Acetylcholinesterase and Its Inhibition**: AChE plays a pivotal role in regulating ACh levels by breaking it down. AChEIs can enhance cholinergic signaling, leading to improved cognitive and muscular functions in various conditions.

5. **Clinical Applications**: AChEIs have been successfully used in treating Alzheimer's disease and other neurological disorders, demonstrating their efficacy in improving symptoms and quality of life for patients.

6. **Safety and Side Effects**: While AChEIs offer therapeutic benefits, they also carry the risk of side effects and toxicity. Monitoring and individualized treatment plans are essential for patient safety.

7. **Natural Compounds and Lifestyle Factors**: Natural AChE inhibitors and dietary influences on ACh synthesis highlight the potential for integrative approaches to support cholinergic function.

8. **Future Directions**: The future of AChEIs lies in innovative drug development, personalized medicine approaches, and a deeper understanding of cholinergic signaling in health and disease.

The Future of Acetylcholine Research

The field of acetylcholine research is rapidly evolving, with several promising directions on the horizon:

1. **Innovative Drug Development**: Ongoing research is focused on developing more selective and effective AChEIs that minimize side effects while maximizing therapeutic benefits. This includes exploring allosteric modulators and dual-action compounds.
2. **Interdisciplinary Collaboration**: Future research will benefit from collaborations across neuroscience, pharmacology, nutrition, and psychology to fully understand ACh's role in health and disease and to develop comprehensive treatment strategies.
3. **Personalized Medicine**: The integration of genetic and biomarker research will pave the way for tailored therapies that optimize treatment for individuals based on their unique biological profiles.
4. **Holistic Approaches**: Exploring the role of lifestyle factors, such as diet and exercise, in modulating ACh levels will enhance the overall management of conditions associated with cholinergic dysfunction.
5. **Expanding Applications**: Beyond neurological disorders, AChEIs may find new applications in areas such as sports medicine and cognitive enhancement, necessitating ethical considerations and rigorous scientific validation.

Final Thoughts

Mastering the complexities of acetylcholine and its inhibitors presents a significant opportunity for advancing therapeutic interventions in a range of health conditions. By understanding and applying the knowledge gained from this book, healthcare professionals and researchers can contribute to improving the lives of those affected by cholinergic dysfunctions.

The journey of research into acetylcholine and its myriad roles is far from complete. As science progresses, so too will our understanding of this vital neurotransmitter and its potential to enhance health and well-being across the lifespan.

Chapter 25: Resources and Further Reading

Introduction

As we conclude our exploration of acetylcholine and its significant role in the nervous system, it is essential to provide readers with resources for further study. This chapter includes recommended books, articles, journals, and online resources that will deepen your understanding of acetylcholine, acetylcholinesterase inhibitors, and their therapeutic implications.

Recommended Books

"Neuroscience: Exploring the Brain" by Mark F. Bear, Barry W. Connors, and Michael A. Paradiso

A comprehensive textbook that covers the fundamentals of neuroscience, including neurotransmission, neural mechanisms, and the role of neurotransmitters like acetylcholine.

"Pharmacology" by Gary C. D. W. McGraw-Hill

This book provides an in-depth look at pharmacological principles, including detailed information on drug interactions, therapeutic uses, and mechanisms of action for various drug classes, including AChE inhibitors.

"Cognitive Neuroscience: The Biology of the Mind" by Gazzaniga, Ivry, and Mangun

A great resource for understanding the relationship between brain function, neurotransmitters, and cognition, focusing on the role of acetylcholine in memory and learning.

"Alzheimer's Disease: A Comprehensive Guide to the Disease and Its Management" by Charles F. Reynolds III

This book offers insights into Alzheimer's disease, highlighting the mechanisms involved, including the cholinergic hypothesis and the role of AChE inhibitors in treatment.

Key Journals

Journal of Neurochemistry

A leading journal that publishes research on neurotransmission, including studies related to acetylcholine and its role in health and disease.

Neuropharmacology

This journal covers the pharmacological aspects of the nervous system, including research on acetylcholinesterase inhibitors and their therapeutic applications.

Alzheimer's Research & Therapy

A specialized journal focusing on research related to Alzheimer's disease and other neurodegenerative disorders, including studies on the efficacy of AChE inhibitors.

Frontiers in Neuroscience

An open-access journal that publishes research across various areas of neuroscience, providing insights into the latest discoveries in neurotransmission and cholinergic systems.

Articles of Interest

"Acetylcholine: A Key Modulator of Neural Activity" - Trends in Neurosciences

This article reviews the functions of acetylcholine in the brain and its implications for cognitive processes and neurodegenerative diseases.

"The Role of Acetylcholinesterase Inhibitors in Neurodegeneration" - Journal of Alzheimer's Disease

A comprehensive overview of how AChE inhibitors are used in the management of Alzheimer's disease and their potential effects on cognitive function.

"Natural Products as Acetylcholinesterase Inhibitors: A Review" - Current Medicinal Chemistry

An exploration of various natural compounds that inhibit AChE, highlighting their sources and potential therapeutic uses.

Online Resources

PubMed

A free resource that provides access to a vast database of biomedical literature, including research articles on acetylcholine, AChE inhibitors, and related topics.

Google Scholar

An academic search engine that helps find scholarly articles, theses, books, and conference papers related to acetylcholine and its role in health and disease.

ClinicalTrials.gov

A database of privately and publicly funded clinical studies conducted around the world, offering insights into ongoing research involving AChE inhibitors and other therapeutic agents.

Alzheimer's Association

A valuable resource for information on Alzheimer's disease, including research updates, treatment options, and support for patients and caregivers.

Conclusion

The study of acetylcholine and acetylcholinesterase inhibitors is a dynamic and evolving field with significant implications for neuroscience and medicine. By utilizing these resources, readers can enhance their knowledge and stay informed about the latest research and developments related to acetylcholine, paving the way for future explorations and applications in both clinical and personal health contexts.

www.ingramcontent.com/pod-product-compliance
Lightning Source LLC
Chambersburg PA
CBHW082113220526
45472CB00009B/2157